U.S. ENVIRONMENTAL PROTECTION AGENCY

OFFICE OF INSPECTOR GENERAL

National Association of State Departments of Agriculture Research Foundation Needs to Comply With Certain Federal Requirements and EPA Award Conditions to Ensure the Success of Pesticide Safety Education Programs

Report No. 14-P-0131   March 10, 2014

Report Contributors:

Angela Bennett
Philip Cleveland
Bill Spinazzola

Abbreviations

ACH	Automated Clearing House
CFR	Code of Federal Regulations
EPA	U.S. Environmental Protection Agency
FFR	Federal Financial Report
FY	Fiscal Year
MTDC	Modified Total Direct Costs
NASDA	National Association of State Departments of Agriculture
NASDARF	National Association of State Departments of Agriculture Research Foundation
NBC	National Business Center
OGD	Office of Grants and Debarment
OIG	Office of Inspector General
OMB	Office of Management and Budget
SF	Standard Form

Cover photo: Pesticide being applied to a field. (EPA photo)

Hotline

To report fraud, waste or abuse, contact
us through one of the following methods:

email: OIG_Hotline@epa.gov
phone: 1-888-546-8740
fax: 1-202-566-2599
online: http://www.epa.gov/oig/hotline.htm

write: EPA Inspector General Hotline
1200 Pennsylvania Avenue, NW
Mailcode 2431T
Washington, DC 20460

Suggestions for Audits or Evaluations

To make suggestions for audits or evaluations,
contact us through one of the following methods:

email: OIG_WEBCOMMENTS@epa.gov
phone: 1-202-566-2391
fax: 1-202-566-2599
online: http://www.epa.gov/oig/contact.html#Full_Info

write: EPA Inspector General
1200 Pennsylvania Avenue, NW
Mailcode 2410T
Washington, DC 20460

At a Glance

Why We Did This Review

The U.S. Environmental Protection Agency (EPA) Office of Inspector General (OIG) reviewed the funds drawn by the National Association of State Departments of Agriculture Research Foundation (NASDARF) under EPA Cooperative Agreement No. 83456201. The award provided $3.6 million to:

- Evaluate and improve existing pesticide safety training programs and materials.
- Identify areas in the program where additional pesticide safety education is needed.

The purpose of our examination was to determine whether costs incurred were allowable in accordance with federal requirements and the agreement and whether NASDARF conducted procurements in accordance with federal regulations and the agreement.

This report addresses the following EPA themes:

- *Embracing EPA as a high performing organization.*
- *Taking action on toxics and chemical safety.*

For further information, contact our public affairs office at (202) 566-2391.

The full report is at:
www.epa.gov/oig/reports/2014/
20140310-14-P-0131.pdf

National Association of State Departments of Agriculture Research Foundation Needs to Comply With Certain Federal Requirements and EPA Award Conditions to Ensure the Success of Pesticide Safety Education Programs

What We Found

NASDARF's financial management system did not meet certain federal requirements and conditions of the EPA award. Specifically, NASDARF incorrectly calculated and applied indirect cost rates, reported

We questioned $571,626 of potentially unallowable costs.

outlays for indirect costs in excess of recorded expenses, and drew funds that exceeded its cash needs. As a result, we questioned $275,650.

NASDARF did not document its procurement selection process or provide documentation to support any cost or price analysis performed on its project management subcontract as required by the Code of Federal Regulations (CFR) in 40 CFR Part 30. NASDARF did not determine the reasonableness of costs for two subgrants as required by conditions of the award. In addition, NASDARF's written procurement policy lacked procedures to ensure compliance with 40 CFR Part 30. As a result, we questioned $295,976.

The OIG also identified an unresolved issue pertaining to potentially unallowable costs of $118,324 drawn under a prior EPA award. The costs, recorded as a refundable advance, represent funds received as of year-end but not yet earned.

Recommendations and Responses

We recommend that the EPA disallow and recover $571,626 pertaining to the financial management and procurement issues. We also recommend that the EPA require NASDARF to recalculate its indirect cost rates to be consistent with 2 CFR Part 230 and establish controls to ensure that its financial management and procurement systems comply with federal requirements and conditions of the award. Further, we recommend that certain special conditions be included for all active and future EPA awards until NASDARF meets all applicable federal financial and procurement requirements. For the $118,324 of potentially unallowable costs, the EPA should review and recover costs determined to be unallowable. NASDARF generally did not agree with the OIG findings and recommendations. The EPA agreed with the OIG recommendations and stated it would work with NASDARF to resolve the issues.

Noteworthy Achievements

In response to OIG finding outlines, NASDARF modified its subcontract for project management services and its written procurement procedures to include OIG-recommended requirements pertaining to 40 CFR Part 30.

March 10, 2014

MEMORANDUM

SUBJECT: National Association of State Departments of Agriculture Research Foundation
Needs to Comply With Certain Federal Requirements and EPA Award Conditions to
Ensure the Success of Pesticide Safety Education Programs
Report No. 14-P-0131

FROM: Arthur A. Elkins Jr.

TO: Howard Corcoran, Director
Office of Grants and Debarment
Office of Administration and Resources Management

This is our report on the subject audit conducted by the Office of Inspector General (OIG) of the
U.S. Environmental Protection Agency (EPA). This report contains findings that describe the problems
the OIG has identified and corrective actions the OIG recommends. This report represents the opinion of
the OIG and does not necessarily represent the final EPA position. Final determinations on matters in
this report will be made by EPA managers in accordance with established audit resolution procedures.

Action Required

In accordance with EPA Manual 2750, you are required to provide us your proposed management
decision on the findings and recommendations contained in this report before you formally complete
resolution with the recipient. Your proposed management decision is due in 120 days, or on July 8, 2014.
To expedite the resolution process, please also email an electronic version of your management decision
to adachi.robert@epa.gov.

Your response will be posted on the OIG's public website, along with our memorandum commenting on
your response. Your response should be provided as an Adobe PDF file that complies with the
accessibility requirements of Section 508 of the Rehabilitation Act of 1973, as amended. The final
response should not contain data that you do not want to be released to the public; if your response
contains such data, you should identify the data for redaction or removal. This report will be available
on our website at http://www.epa.gov/oig.

If you or your staff have any questions regarding this report, please contact Richard Eyermann,
acting Assistant Inspector General for Audit, at (202) 566-0565 or eyermann.richard@epa.gov; or
Robert Adachi, Product Line Director, at (415) 947-4537 or adachi.robert@epa.gov.

National Association of State Departments of Agriculture
Research Foundation Needs to Comply With Certain
Federal Requirements and EPA Award Conditions to Ensure
the Success of Pesticide Safety Education Programs

14-P-0131

Table of Contents

Chapters

-continued-

National Association of State Departments of Agriculture
Research Foundation Needs to Comply With Certain
Federal Requirements and EPA Award Conditions to Ensure
the Success of Pesticide Safety Education Programs

14-P-0131

Appendices

Chapter 1
Independent Accountant's Report

As part of our oversight of assistance agreement awards made by the
U.S. Environmental Protection Agency (EPA), the Office of Inspector General (OIG)
examined the costs claimed under Cooperative Agreement No. 83456201 awarded to
the National Association of State Departments of Agriculture Research Foundation
(NASDARF). The OIG conducted the examination to determine whether the costs
incurred were allowable under federal regulations and conditions of the award, and
whether procurements were conducted in accordance with federal regulations. The
applicable federal requirements found in the Code of Federal Regulations (CFR)
include:

- Title 2 CFR Part 230, *Cost Principles for Non-Profit Organizations.*
- Title 40 CFR Part 30, *Uniform Administrative Requirements for Grants
  and Agreements with Institutions of Higher Education, Hospitals, and
  Other Non-Profit Organizations.*

By accepting funding provided through the agreement, NASDARF has responsibility
for complying with these requirements. Our responsibility is to express an opinion on
NASDARF's compliance based on our examination.

We conducted our examination in accordance with generally accepted government
auditing standards issued by the Comptroller General of the United States. We also
utilized the attestation standards established by the American Institute of Certified
Public Accountants. We examined, on a test basis, evidence supporting NASDARF's
assertion contained in its interim Federal Financial Report (FFR) for the period ending
January 24, 2013, and performed other procedures as we considered necessary.
We believe that our examination provides a reasonable basis for our opinion.

We interviewed EPA personnel from the Office of Grants and Debarment and the
Office of Pesticide Programs in headquarters, Washington, D.C. We obtained an
understanding of the agreement and gathered information concerning NASDARF's
performance. Specifically, we reviewed NASDARF's request for proposal to determine
the objectives of the agreement and project deliverables. We also gathered information
on criteria relevant to the agreement and reviewed applicable federal requirements,
including 2 CFR Part 230 and 40 CFR Part 30.

On January 10, 2013, we conducted an entrance conference with NASDARF.
We returned to NASDARF's office in Washington, D.C., the week of January 22,
2013, to conduct interviews and obtain documentation to address our objectives.

To determine whether the costs incurred by NASDARF were allowable under 40 CFR Part 30 and conditions of the agreement, we:

- Conducted interviews to obtain an understanding of the accounting system, applicable internal controls, and practices followed to administer the agreement.
- Reviewed single audit reports for fiscal years (FYs) 2010 through 2012 to identify issues that may have an impact on our examination.
- Reviewed internal controls related to the assignment objectives and performed tests to determine whether the controls are in place and operating effectively.
- Reviewed compliance with laws, regulations and the terms and conditions of the agreement.
- Examined reported outlays on a test basis to determine whether the outlays were adequately supported and eligible for reimbursement under the conditions of the agreement and federal requirements.
- Identified required deliverables and determined whether they were submitted to and accepted by the EPA.

To determine whether NASDARF conducted its procurements in accordance with 40 CFR Part 30, we:

- Conducted interviews to obtain an understanding of how NASDARF procured subcontracts and subgrants.
- Obtained and reviewed written procurement and contract administration policies and procedures to determine whether they included federal requirements.
- Obtained and reviewed supporting documentation for the award of the sole-source subcontract for project management services to ensure compliance with relevant criteria.
- Obtained and reviewed supporting documentation for a judgmental sample of three of eight subgrant and subcontract awards to ensure compliance with relevant criteria. The sample represented 81% ($289,943 of $357,995) of the total subgrant and subcontract costs.

We also reviewed project costs and NASDARF's drawdown of EPA funds. Specifically, we performed the following steps:

- Obtained, reviewed and reconciled NASDARF's interim FFR for the period ending January 24, 2013.
- Discussed the FFR preparation with NASDARF to ensure the FFR was prepared in accordance with applicable laws, regulations, and terms and conditions of the cooperative agreement.
- Reviewed NASDARF's drawdown procedures, obtained a draw history, and selected a judgmental sample of three of the 15 draws to determine whether the draws were reasonable and properly supported. The sample

represented 24% ($350,000 of $1,460,000) of total draws as of December 6, 2012. We reviewed supporting invoices, payment documents and associated accounting system entries to determine whether the expenditures were allocable and allowable under 40 CFR Part 30 and the agreement.

- Obtained and analyzed monthly bank statements to determine compliance with cash management requirements as described in the administrative conditions of the agreement and under federal regulation.

We conducted our examination from January 10, 2013 through November 4, 2013. Our examination disclosed material noncompliance and internal control weaknesses with financial management and procurement. In particular, NASDARF's:

- Financial management system pertaining to cash draws and project costs does not meet the requirements of the agreement or 2 CFR Part 230 and 40 CFR Part 30. Chapter 4 of this report includes a discussion of the noncompliance.

- Subcontract procurement of its project manager was not documented and did not include cost and price analysis as required by the agreement or 40 CFR Part 30. In addition, the subgrant procurements examined did not comply with the agreement's administrative condition no. 4(a). Chapter 5 of this report presents more details on these conditions.

Unless NASDARF can demonstrate compliance with all applicable requirements, we recommend that the EPA disallow and recover $492,817 of reported costs and $78,809 in excess cash draws.

In our opinion, because of the effect of the issues described above, the costs claimed do not meet, in all material respects, the requirements of 2 CFR Part 230 and 40 CFR Part 30, or the conditions of the agreement for the period ended December 31, 2012.

Robert K. Adachi

Robert K. Adachi
Director for Forensic Audits
March 10, 2014

Chapter 2
Introduction

Purpose

The OIG conducted this examination to determine whether NASDARF complied with federal requirements and the conditions of EPA Agreement No. 83456201. Our objectives were to determine whether:

- Costs incurred by NASDARF were allowable in accordance with federal requirements and conditions of the agreement.
- NASDARF followed applicable regulations and conditions of the agreement in its procurement of subcontracts and subgrants.

Background

The National Association of State Departments of Agriculture (NASDA) is a nonprofit, nonpartisan association of public officials comprised of executive heads from 50 state departments of agriculture, as well as executives from the U.S. territories of Puerto Rico, Guam, American Samoa and the U.S. Virgin Islands. NASDA's mission is to support and promote the American agricultural industry through the development, implementation and communication of sound public policy programs. NASDA organized NASDARF to design and implement educational programs relating to farming and other agricultural activities, and to help state and local government develop government programs relating to agricultural activities.

The EPA awarded Agreement No. 83456201 to NASDARF on March 24, 2010. The award provided $3.6 million to support national and international pesticide safety education programs designed to reduce the pesticide exposure of agricultural workers and pesticide applicators. Pesticide safety education projects not only target workers and applicators, but also growers, healthcare providers, and pesticide producers and retailers. The purpose of the award was to evaluate and improve existing pesticide safety training programs and materials, and identify areas within the programs where additional pesticide safety education is needed.

The EPA awarded the funds based on the authorities provided by:

- The Federal Insecticide, Fungicide, and Rodenticide Act, Section 20, which authorizes the agency to issue assistance agreements for research, development, monitoring, public education, training, demonstrations and studies concerning pesticide-related matters.

- The National Environmental Policy Act, Section 102(2)(F), which recognizes the worldwide and long-range character of environmental problems; and, where consistent with U.S. foreign policy, lends appropriate support to programs designed to maximize international cooperation in anticipating and preventing a decline in the world environment.

The budget and project period for Agreement No. 83456201 is April 5, 2010, through April 6, 2015. This agreement is a continuation of work started under NASDARF's previous EPA Agreement No. 83235401.

Although NASDARF is not required to submit a standard form (SF) 425 FFR until the agreement is completed, NASDARF prepared an interim FFR at the OIG's request. The interim FFR, dated January 24, 2013, covered the period April 5, 2010 through December 31, 2012. The FFR included a federal share of outlays totaling $1,398,210, cash receipts of $1,460,000 and cash on hand of $78,809.

Noteworthy Achievements

On June 19, 2013, the OIG issued finding outlines to NASDARF. The outlines expressed concern that NASDARF's project management subcontract did not include 40 CFR §30.48(c)(1) and (d) requirements pertaining to breach of contract remedies and access to records. The finding outlines included a recommendation that NASDARF revise the subcontract to include these requirements. In response, NASDARF stated that it had modified the subcontract to reflect the requested adjustments. The OIG reviewed the revised subcontract and confirmed the provisions were included. NASDARDF also revised its written procurement procedures to include OIG-recommended requirements pertaining to 40 CFR Part 30.

NASDARF and EPA Responses

OIG received comments from NASDARF and the EPA concerning the OIG's draft report issued on November 5, 2013. NASDARF also provided supplemental documentation to support some comments. Due to the various types and volume of documents provided, the OIG did not include NASDARF's additional documentation in this report, but the information is available upon request.

We held an exit conference with NASDARF on February 12, 2014 to discuss their response to the draft report and the impact on our final report. NASDARF continued to disagree with our findings.

The OIG did not hold an exit conference with EPA. Rather EPA requested that the OIG proceed with issuing the audit report so they can begin working with NASDARF to resolve the findings.

Chapter 3
Results of Examination

NASDARF's procurement practices, methods for determining and claiming indirect costs, and drawdown practices for EPA Agreement No. 83456201 did not comply with certain federal requirements or agreement conditions. As shown in table 1, we questioned $571,626 of reported costs.

Table 1: Costs reported and questioned for EPA Agreement No. 83456201

Cost element	Reported	Questioned	Note
Travel	$ 382,417		
Supplies	24,618		
Contractual	611,287	295,976	(2)
Other	183,047		
Indirect costs	196,841	196,841	(3)
Subtotal (1)	$1,398,210	$492,817	
Drawdown amounts	$1,460,000	78,809	(4)
Total		$571,626	

Source: OIG-generated table.

Note 1: Represents the amount reported (including expenditures and unliquidated obligations) on the interim SF 425 FFR prepared by NASDARF and provided to the OIG on January 24, 2013.

Note 2: Contractual costs totaling $295,976 were questioned, including:

- Costs of $151,484 for Ramsey Consulting because NASDARF did not perform a cost or price analysis, justify the use of a sole source, or document the basis for the award cost or price of the sole-source award. Chapter 5 of this report discusses this issue in detail.

- Costs of $144,492 ($44,492 for University of Florida and $100,000 for CropLife Latin America), because NASDARF did not comply with agreement condition 5a(4) pertaining to the procurement of subgrants. The agreement condition requires that the recipient determine the reasonableness of proposed subgrant costs. NASDARF was unable to provide documentation that the costs were reasonable. Chapter 5 of this report discusses this issue in detail.

Note 3: We questioned $196,841 of indirect costs because NASDARF did not comply with the requirements of its indirect cost rate agreements when calculating an indirect cost rate and applying the rate to the allocation base for the agreement. The amount questioned also includes a negative amount of $12,896 for costs related to two unsupported adjustments to indirect costs. Chapter 4 of this report discusses these issues in detail.

Note 4: We questioned the cash-on-hand balance of $78,809, which NASDARF reported on its interim SF 425, because NASDARF maintained excessive cash balances under the agreement and did not use the funds in a timely manner. This amount represents the difference in expenditures and receipts. Chapter 4 of this report discusses the issue in detail.

Recommendation

We recommend that the Director of the Office of Grants and Debarment:

1. Disallow and recover $571,626 of questioned costs. If NASDARF provides documentation that meets appropriate federal requirements or demonstrates the fairness and reasonableness of the subcontract and subgrant costs, the amount to be recovered may be adjusted accordingly.

NASDARF and EPA Responses

NASDARF generally disagreed with our findings and provided a detailed response to the issues presented in Notes 2, 3, and 4 above. However, NASDARF did not provide a response to recommendation 1 or the recommendations contained in chapters 4 and 5. The OIG included a discussion of NASDARF's responses to Notes 2, 3, and 4 and OIG comments in the applicable chapters of this report. NASDARF's complete written response and additional OIG comments are included in appendix A.

The EPA agreed with recommendation 1 and stated it will:

1. Provide NASDARF the opportunity to submit documentation to substantiate the questioned costs.
2. Review the documentation and take necessary corrective action, including recovery of all or part of the questioned subcontract and indirect costs as well as funds drawn.
3. Work with NASDARF to implement corrective actions to comply with federal requirements on assuring the reasonableness of sub-grant, sub-contracts, indirect costs and drawdown amounts.

The EPA's complete written response is included in appendix B.

OIG Comments

The OIG agrees with EPA's intended corrective actions for recommendation 1. However, because the EPA did not provide specific planned completion dates for the corrective actions, the OIG considers the status of the recommendation unresolved.

Chapter 4
Financial Management System Did Not Meet Certain Federal Requirements

NASDARF's financial management system did not meet certain federal regulations and agreement conditions. Specifically, NASDARF:

- Did not calculate and apply indirect cost rates correctly.
- Reported outlays for indirect costs in excess of recorded expenses in the general ledger.
- Drew down funds that exceeded cash needs.

These conditions occurred because NASDARF did not have controls in place to ensure compliance with federal regulations and agreement conditions. In calculating indirect cost rates, NASDARF was not aware initially of 2 CFR Part 230 requirements to exclude subgrants and subcontracts in excess of $25,000. NASDARF later interpreted the requirements erroneously and only excluded amounts for subgrants but not subcontracts. For the reported outlays, NASDARF lacked controls to ensure the recording of adjustments in the general ledger. Lastly, NASDARF's written procedures for drawdowns did not incorporate cash-management requirements as described in the administrative conditions of the agreement or under federal regulations.

As described in chapter 3, we questioned $196,841 of indirect costs and $78,809 related to the excessive cash draws.

Indirect Cost Rates Were Calculated and Applied Incorrectly

NASDARF did not calculate indirect cost rates in accordance with the requirements of its approved indirect cost rate agreements. Additionally, NASDARF did not apply the rates to the appropriate allocation base when reporting costs under the agreement.

The U.S. Department of the Interior's National Business Center (NBC) negotiates indirect cost rates for the EPA. The indirect rate agreement between NASDARF and the NBC for the fiscal year ending June 30, 2011, states that the allocation base for the indirect rate should be total direct costs less capital expenditures (i.e., the portion of subgrants or subcontracts in excess of the first $25,000 and pass-through funds). This is consistent with 2 CFR Part 230, Appendix A, paragraph D.3.f., which references modified total direct costs (MTDC) and states, in part:

> Indirect costs shall be distributed to applicable sponsored awards and other benefiting activities within each major function on the

basis of MTDC. MTDC consists of all salaries and wages, fringe benefits, materials and supplies, services, travel, and subgrants and subcontracts up to the first $25,000 of each subgrant or subcontract (regardless of the period covered by the subgrant or subcontract).

Calculation of Indirect Cost Rate

When calculating the allocation base for FY 2011, NASDARF did not exclude any subgrant or subcontract amounts. Subcontracts and subgrants in the allocation base for the EPA agreement amounted to $306,563. The OIG's analysis revealed that four subcontractors or subgrantors received payments that exceeded the $25,000 limit. NASDARF should have excluded the excess payments totaling $86,399 from the allocation base.

We do not know the effect that this condition has on the indirect cost rate. The allocation base consisted of costs from five programs. Our review included only the costs in the base for the EPA agreement. We did not determine whether NASDARF should have deleted any costs from the other four programs. We reviewed the indirect rate calculation for FY 2011 because, at the time of our audit, this was the latest year for which NASDARF had an approved final indirect cost rate.

Application of Indirect Cost Rate

When applying the indirect cost rate to total costs incurred for the EPA agreement, NASDARF excluded $66,758 for subgrant and subcontract payments that exceeded $25,000. We reviewed subgrant and subcontract payments and determined that $286,453 should have been excluded—an increase of $219,695 over the excluded amount.

Initially, NASDARF was not aware of the requirement to exclude amounts in excess of $25,000. Later, NASDARF misinterpreted the requirements and did not exclude amounts for the subcontract that it awarded for project management. NASDARF stated it did not consider the project manager to be the same as other awards. NASDARF also stated that in order for the award to be financially viable, it would need to apply indirect cost rates to the full amount of fees paid.

NASDARF reported $196,841 of indirect costs under the agreement. We questioned the full amount because NASDARF did not calculate and apply the rates in accordance with the provisions of its negotiated indirect cost rate agreements. To ensure that its indirect cost rates result in an equitable allocation of costs, NASDARF should:

- Recalculate indirect cost rates after excluding subgrant and subcontract amounts in excess of $25,000, or request the NBC to amend its indirect cost rate agreements to include an equitable allocation base.

- Submit revised indirect cost rates to the NBC for approval.
- Claim indirect costs using the revised approved rates.

Reported Costs Exceeded Amounts Recorded in General Ledger

NASDARF was unable to support adjustments made to recorded expenses when preparing its FFR. Total outlays reported on the interim FFR were $1,398,210. The reported costs exceeded amounts recorded in the general ledger by $4,123. NASDARF stated the difference was attributable to three adjustments related to indirect costs. NASDARF provided adequate support for unrecorded indirect costs of $17,019. However, we did not receive sufficient support for the two remaining adjustments. NASDARF stated it had not recorded the amounts in the general ledger as expenditures.

One negative adjustment of $5,183 was to reduce indirect expenses for the amount of the CropLife Latin America contract, which exceeded $25,000. The remaining negative adjustment of $7,713 was a decrease to FY 2011 indirect costs for the difference between the billing and final indirect cost rates. Documents that NASDARF provided to support these two adjustments consisted of two summary accounting reports with handwritten notes in the margin. Supporting documents that show how the amounts were determined were not included.

The requirements of 40 CFR §30.21(b)(7) state that recipients' financial management systems shall provide accounting records, including cost accounting records that are supported by source documentation. In addition, 2 CFR Part 230, Appendix A, paragraph A.2g, states that to be allowable under an award, costs must be "adequately" documented. As a result, we questioned the two negative adjustments totaling $12,896. Table 2 shows NASDARF's reconciliation of the FFR to the general ledger.

Table 2: Reconciliation of FFR to the general ledger

		Amounts
Comparison		
FFR line 10g (total federal share)		$ 1,398,210
General ledger report		1,394,087
Difference		$ 4,123
Explanation		
Unrecorded indirect costs for December 2012		$ 17,019
Adjustment to indirect cost for CropLife contract exceeding $25,000	($ 5,183)	
Unrecorded adjustment to FY 2011 indirect costs	(7,713)	
Subtotal adjustments		(12,896)
Difference		$ 4,123

Source: Based on information provided by NASDARF.

Accepted Cash Management Practices Not Followed

NASDARF did not follow accepted cash management practices when drawing down EPA funds. This resulted in funds being held for excessive periods, unnecessary funds drawn, and funds drawn and used for expenses incurred under non-EPA programs. As of December 6, 2012, NASDARF made 15 draws totaling $1,460,000. Our review of three of the 15 draws showed that NASDARF:

- Expended funds from 21 to 125 days after drawing the funds.
- Maintained an average cash balance of $147,029 during the period of July 2010 through February 2013.
- Used EPA funds to pay expenditures for five other non-EPA programs.

This occurred because NASDARF's written drawdown policy does not include requirements to minimize the time elapsing between drawing funds and making expenditures, as required by the agreement and 40 CFR §30.22(a) and (b).

In chapter 3 of this report, we questioned $78,809 of excess funds drawn as of the interim FFR for the period ending January 24, 2013. This amount is reported as cash on hand (line 10c) and represents the excess of cash receipts over cash disbursements. NASDARF did not earn interest on the excess cash funds drawn.

Agreement administrative condition 4(a) states:

> By accepting this agreement for the electronic method of payment through the Automated Clearing House (ACH) network using the EPA-ACH payment system, the recipient agrees to request funds based on the recipient's immediate disbursement requirements by presenting an EPA-ACH Payment Request to your EPA Servicing Finance Office. Further, failure on the part of the recipient to comply with the above conditions may cause the recipient to be placed on the reimbursement payment method.

Title 40 CFR §30.22(a) states, in part:

> Payment methods shall minimize the time elapsing between the transfer of funds from the United States Treasury and the issuance or redemption of checks, warrants, or payment by other means by the recipients.

Title 40 CFR §30.22(b) states, in part:

> Cash advances to a recipient organization shall be limited to the minimum amounts needed and be timed to be in accordance with the actual, immediate cash requirements of the recipient organization in carrying out the purpose of the approved program

or project. The timing and amount of cash advances shall be as close as is administratively feasible to the actual disbursements by the recipient organization for direct program or project costs and the proportionate share of any allowable indirect costs.

Funds Held for Excessive Periods

NASDARF drew funds in advance of expenditures and held the funds for excessive periods. We selected and reviewed supporting documentation for three of the 15 draws. As shown in table 3, NASDARF did not begin to expend funds for 21 to 125 days after the funds were drawn. In addition, the number of days between the draw and the last payment ranged from 110 to 230 days.

Table 3: Number of days to expend drawdowns

Draw date	First payment	Days to first payment	Last payment	Days to last payment
7/21/2010	8/11/2010	21	3/8/2011	230
7/21/2011	8/22/2011	32	11/08/2011	110
7/12/2012	11/14/2012	125	1/22/2013	194

Source: NASDARF drawdown requests and supporting documents.

We noted a fourth draw made on April 22, 2013, for $249,000. In response to our request for supporting documentation for the funds drawn, NASDARF replied that the $249,000 represents expenditures that it expects to incur in April, May and June 2013. NASDARF provided the OIG with a listing of the expected expenditures for that 3-month period, which totaled $248,585. Since the information provided represents expected expenditures, we were unable to determine if the information refers to the number of days to expend the draw or if the draw was appropriate.

Excessive Cash Balances Maintained

NASDARF maintained excessive cash balances in its EPA bank account. We reviewed bank statements for the period beginning with the first EPA draw of funds in July 2010 through February 2013. We summarized deposits, checks and other debits, transfers in, and transfers out. We utilized the summary to compute an average cash balance for each month and fiscal year, and for the agreement period July 2010 through February 2013 (32 months). We found that NASDARF maintained large cash balances during the entire period. As shown in table 4, the average cash balance by year ranged from a low of $43,472 in FY 2011 to a high of $263,164 in FY 2012.

Table 4: Average cash balance

Fiscal year	Average cash balance	Lowest balance	Highest balance
2011	$119,900	$43,472	$233,651
2012	$185,748	$108,060	$263,164
2013	$129,642	$66,380	$175,409

Source: OIG calculations using NASDARF bank statements.

Funds Used for Other Purposes

NASDARF used EPA funds to cover expenditures under other non-EPA programs. Our review of NASDARF's bank statements showed numerous transfers of funds from the EPA bank account to five other NASDARF bank accounts, including accounts for:

- NASDA.
- Agricultural Quality Inspection Services.
- Midwestern Association of State Departments of Agriculture.
- Northeastern Association of State Departments of Agriculture.
- Western Association of State Departments of Agriculture.

NASDARF stated it used the EPA funds to pay expenses incurred on other non-EPA programs. This occurred because of delays in receiving funds from other entities, delays due to the availability of funds by the bank, and problems with NASDARF's bill-paying service. Table 5 summarizes transfers to and from the EPA bank account.

Table 5: Transfers to and from the EPA account

	Transfers out	Transfers in
FY 2010 (April–June 2010)	0	0
FY 2011 (July 2010–June 2011)	$ 35,371	$126,753
FY 2012 (July 2011–June 2012)	244,951	11,698
FY 2013 (July 2012–June 2013)	53,855	0

Source: NASDARF monthly bank statements.

Accounting Policy Manual Does Not Include Required Procedures

NASDARF's draft accounting policy manual dated June 30, 2011, *Internal Control Memo,* includes procedures for drawing federal funds. However, the procedures only indicate who requests the draws and states that the automated accounting software be used to summarize and analyze documentation. There were no written procedures to minimize the time between transfer of funds from the U.S. Treasury and payment by the recipient for program purposes, as required by the agreement and federal regulations.

Title 40 CFR §30.21(b)(5) states, in part:

> (b) Recipients' financial management systems shall provide for the following.

> (5) Written procedures to minimize the time elapsing between the transfer of funds to the recipient from the U.S. Treasury and the issuance or redemption of checks, warrants or payments by other means for program purposes by the recipient.

Conclusion

Based on the findings above, NASDARF's financial management system did not meet certain federal requirements contained in the agreement. The recipient's written policies and procedures do not include the necessary guidance to ensure compliance with 40 CFR Part 30. We have no assurance that costs are fair and reasonable when NASDARF calculates indirect costs and rates incorrectly, does not reconcile reported costs to its general ledger, withdraws excessive cash, and holds funds for a long period of time.

To address the financial management issues identified, NASDARF should establish controls to ensure indirect costs are calculated and applied consistent with the requirements found in 2 CFR Part 230. NASDARF should also ensure that costs are recorded and supported by adequate source documentation as required by 40 CFR §30.21(b)(7), and ensure that drawdowns are consistent with the requirements contained in 40 CFR §30.21(b)(5) and §30.22(a) and (b).

The EPA should also impose special conditions on all current and future awards of EPA funds as outlined in 40 CFR Part 30 and 40 CFR §30.14, *Special Award Conditions*, which states:

> If an applicant or recipient: has a history of poor performance, is not financially stable, has a management system that does not meet the standards prescribed in Circular A-110; has not conformed to the terms and conditions of a previous award; or is not otherwise responsible, EPA may impose additional requirements as needed, provided that such applicant or recipient is notified in writing as to: the nature of the additional requirements, the reason why the additional requirements are being imposed, the nature of the corrective action needed, the time allowed for completing the corrective actions, and the method for requesting reconsideration of the additional requirements imposed. Any special conditions shall be promptly removed once the conditions that prompted them have been corrected.

The special conditions should include: (a) payment on a reimbursement basis; and (b) EPA review and approval of reimbursement requests prior to payment.

Recommendations

We recommend that the Director of the Office of Grants and Debarment:

2. Require NASDARF to:

   a. Recalculate its indirect cost rate for FY 2011, excluding subgrant and subcontract amounts in excess of $25,000; and submit to the NBC for approval, or ask the NBC to amend, the indirect cost rate agreements to include an equitable allocation base.
   b. Claim indirect costs using the recalculated approved rates.

3. Require NASDARF to calculate its indirect cost rates for years beyond FY 2011 by excluding subgrant and subcontract amounts in excess of $25,000; or to be in accordance with any revised indirect cost rate agreement.

4. Require the following special conditions be included for all current and future awards until the EPA determines that NASDARF has met all applicable federal financial requirements:

   a. Payment on a reimbursement basis.
   b. Review and approval of reimbursement requests, including all supporting documentation, prior to payment.

NASDARF Response

NASDARF generally did not agree with the OIG findings and recommendations and provided the following comments.

Did not calculate and apply indirect cost rates correctly: NASDARF responded that it is fully informed of 2 CFR Part 230 requirements and at no time was under the impression, nor did they execute any activities under the illusion, that indirect costs could be charged on subgrants and subcontracts in excess of $25,000. Further, NASDARF stated that the OIG's charge relates specifically to Ms. Carol (Ramsay) Black's misclassification as a "subcontractor," when in fact, Ms. Black is a contracted employee/consultant under Ramsay Consulting, LLC. As a contracted employee, Ms. Black is not subject to the rule of a subcontract under 2 CFR Part 230.

NASDARF also responded that the OIG listed Ramsay Consulting LLC, Richard Herrett, Marla Stein Associates, and CropLife Latin America as subcontractors/subgrantors receiving payments in excess of $25,000. Ramsay

Consulting, LLC and Mr. Herrett served as project managers for the grant and are therefore considered direct salary. Irrespective of the classification, Mr. Herrett was not paid under this current agreement. Marla Stein Associates received $21,143 under Cooperative Agreement 83456201 and the remaining invoices were charged under a prior agreement.

NASDARF stated it is awaiting the EPA grants office's opinion before making any decisions in regard to reapplying for indirect rates.

<u>NASDARF lacked controls to ensure the recoding of adjustment in the general ledger</u>: NASDARF disagreed with the OIG's finding that it lacked controls to ensure the recording of adjustments in the general ledger. NASDARF stated that the nature of any indirect rate application results in regular adjustments when rates change and new payments are made for subgrants. NASDARF's goal is to obtain a correct cumulative total in the indirect allocation, not an unrealistic expectation of an unchanging allocation each month. Assessing these values at interim periods over the life of the grant only offers a snapshot at that specific time. NASDARF also provided a revised SF 425 as of October 31, 2013, and supporting general ledger through October 2013.

<u>NASDARF's written procedures for drawdown did not incorporate cash-management requirements described in the administrative conditions of the agreement or under federal regulations</u>: NASDARF responded it had changed both personnel and policy within the last year to ensure cash management practices are line with grant requirements. NASDARF's updated policy now reads:

> Drawdowns related to the EPA agreement are initiated by the Director of Finance based on anticipated short-term expenses as reported and documented by the Project Director. Drawdowns are typically made after expenses are paid but may be made a week in advance in anticipation of large vendor/subgrant disbursements.

NASDARF also stated that it is aware of OIG's position claiming funds are available within 3 business days; however, recent technical difficulties with the ASAP website make it extremely challenging for NASDARF management to operate with a 3-day window for distributing funds.

<u>NASDARF used EPA funds to cover expenditures under other non-EPA programs</u>: NASDARF stated that the OIG's assertion that it used EPA funds for expenditures for other non-EPA programs is incorrect. NASDASRF stated that it transfers funds from the NASDARF checking account to other NASDA-affiliated accounts only after NASDARF earned the funds through indirect allocations or reimbursement for direct expenses.

EPA Response

The EPA agreed with the OIG's recommendations and provided the following comments.

Recommendation 2: The Office of Grants and Debarment (OGD) will provide NASDARF the opportunity to clarify its position and properly support its indirect rate and require NASDARF to amend the rate where appropriate. OGD will review NASDARF's support and coordinate with the U.S. Department of the Interior's NBC to review and amend the indirect cost rate agreement as needed, then require NASDARF to adjust claimed costs according to the revised agreement.

Recommendation 3: Where necessary, OGD will require NASDARF to recalculate its indirect costs rates to be in compliance with 2 CFR Part 230 and coordinate with NBC to revise NASDARF's rates. OGD will then require NASDARF to adjust claimed costs according to any revised indirect cost rate agreements.

Recommendation 4: OGD will place NASDARF on reimbursement for its active assistance agreements. The agency will require NASDARF to provide supporting documents for costs incurred for the agency's review prior to releasing funds for payment. NASDARF will remain on reimbursement status until it has demonstrated to the agency that its financial management of EPA agreements meets applicable federal requirements for drawing grant funds.

OIG Comments

The OIG did not agree with NASDARF's comments. The OIG's responds to them as follows.

Did Not Calculate and Apply Indirect Cost Rates Correctly: In its response, NASDARF states that Ms. Black's position should be treated as direct salary. The OIG disagrees with this position. Under 2 CFR Part 230, Appendix B.8, compensation for personal services includes services provided by the organization's employees. Since Ms. Black is not an employee, her position cannot be treated as a direct salary. There is no disagreement that Ms. Black is a contractor. However, based on further review of 2 CFR Part 230, the OIG believes that clarification is needed about the relationships discussed (i.e., grants-subgrants, contracts-subcontracts, and whether the requirement applies to the Ramsey contract). NASDARF's relationship with Ramsey is not the typical grant-subgrant relationship; rather, it is a grant-contract relationship. We recommend that the EPA seek clarification on whether this type of relationship is subject to the $25,000 requirement. This would also apply to the services provided by Mr. Herrett.

NASDARF's comment that the costs for Mr. Herrett and Ms. Stein were not claimed under this agreement is correct. However, the costs were included in the allocation base when computing the FY 2011 indirect cost rate. NASDARF used the FY 2011 rate to claim indirect costs under this agreement. Therefore, these costs should be limited to the $25,000.

NASDARF Lacked Controls to Ensure the Recoding of Adjustment in the General Ledger: The OIG based its finding on the fact that the general ledger did not support the outlays reported on the interim FFR. The OIG acknowledges the adjustments of indirect costs as NASDARF discussed. However, outlays reported on the FFR should be adequately supported as required by 40 CFR §30.21(b)(7). Although NASDARF provided a revised SF 425 and general ledger, it did not provide any additional documentation to support the two adjustments reported under the interim FFR. As a result, the OIG continues to question the costs as unsupported.

NASDARF's Written Procedures for Drawdown Did Not Incorporate Cash-Management Requirements Described in the Administrative Conditions of the Agreement or Under Federal Regulations: The OIG acknowledges that NASDARF revised its procedures. However, the procedures do not indicate that drawdowns should be made as close as administratively feasible to the actual disbursements and be timed in accordance with the actual, immediate cash requirements. The OIG does not consider this acceptable, and continues to maintain that NASDARF can get funds from the EPA within 3 days or less.

NASDARF Used EPA Funds to Cover Expenditures Under Other Non-EPA Programs: The OIG disagrees with NASDARF that it did not use EPA funds for expenditures for non-EPA programs. The OIG's cash flow analysis, conducted using NASDARF's bank statements, showed that NASDARF moved funds between EPA's programs and others. Most of the transfers out were to the NASDA account, which, as discussed in NASDARF's response above, could be for reimbursement for indirect expenses. However, there were significant deposits to and from the EPA account from other agencies. There were also smaller transfers out from the EPA account to other associations within the state departments of agriculture. Federal Regulations at Title 40 CFR 30.21(b)(3) require that recipients' financial management systems shall assure that funds are used solely for authorized purposes.

The OIG agrees with EPA's intended corrective actions for recommendations 2, 3, and 4. However, because the EPA did not provide specific planned completion dates for the corrective actions, the OIG considers the status of the recommendations unresolved.

Chapter 5
Procurements Did Not Meet Federal Requirements or Agreement Conditions

NASDARF's procurement practices for subcontracts and subgrants did not comply with federal requirements or agreement conditions. In particular, NASDARF did not:

- Provide documentation to support the sole-source procurement process used for its project management subcontract, or conduct a cost or price analysis as required by 40 CFR Part 30 and NASDARF's own procurement policy.
- Comply with agreement administrative condition 15.a.(4) for subgrants awarded to the University of Florida and CropLife Latin America.
- Include provisions required by 40 CFR Part 30 in its written procurement policy.

These conditions occurred because of NASDARF's lack of knowledge or a lack of understanding about agreement conditions and federal requirements. Additionally, NASDARF's written policies and procedures provided inadequate guidance for subcontract and subgrant awards. As shown in chapter 3, we questioned $295,976 that NASDARF incurred and reported for three awards. Table 6 details the subcontract and subgrant awards questioned.

Table 6: Subcontract and subgrant costs questioned

Cost element	Reported and questioned
Project management subcontract	$151,484
University of Florida subgrant	44,492
CropLife Latin America subgrant	100,000
Total	$295,976

Source: NASDARF's reported costs.

Sole-Source Procurement Not Justified; Cost or Price Analysis Not Conducted

When awarding a subcontract for project management, NASDARF did not promote open and free competition, perform a cost or price analysis, document the basis for the award cost or price, or properly document its procurement practices. NASDARF signed the subcontract on February 26, 2009. The subcontract provided project management services at an hourly rate of $90. The hourly rate included salary of $74.50 and indirect costs of $15.50.

As of January 2013, NASDARF had incurred and reported $151,484 for this subcontract.

Requirements listed under 40 CFR §30.43 state that all procurement transactions be conducted in a manner to provide, to the maximum extent practical, open and free competition. In addition, 40 CFR §30.45 requires that some form of cost or price analysis be made and documented in the procurement files in connection with every procurement action. According to 40 CFR §30.46, procurement records and files for purchases in excess of the small purchase threshold must include at least the basis for contractor selection, justification for lack of competition when competitive bids or offers are not obtained, and the basis for award cost or price.

NASDARF's own written procurement policy for purchases exceeding $25,000 requires that a cost analysis be conducted and documented in conjunction with every purchase. Justification for the lack of competition should also be documented if competitive bids or offers are not obtained.

NASDARF initially informed us that it did not solicit proposals for project managers from other subcontractors. A subcontractor was selected based on NASDARF's experiences and NASDARF's discussion with other knowledgeable personnel, including EPA staff. We asked NASDARF to provide justification for the lack of competition, the use of a sole source, the basis for contractor selection, the basis for the award price, and its cost or price analysis for the subcontract. NASDARF was unable to provide any of the required documents or provide evidence that it had performed any of the required steps. NASDARF stated that most of their procurement actions for this subcontract were verbal and were not documented.

Subgrant Procurements Did Not Comply With Administrative Condition

NASDARF did not comply with administrative condition 15.a.(4) in the award of subgrants to the University of Florida and to CropLife Latin America. Administrative condition 15.a.(4) required NASDARF to determine whether proposed subgrant costs are reasonable. NASDARF's analysis of proposed costs for these two awards was insufficient and lacked numerous support items that should be included when determining the reasonableness of costs.

University of Florida Agreement

NASDARF made an $88,984 award to the University of Florida on May 25, 2010. The purpose of the award was to identify the potential risks of pesticide drift from agricultural operations, develop good-neighbor practices for minimizing risks, and disseminate educational material on drift and good-neighbor practices.

NASDARF agreed to pay $44,492 upon the execution of the award, with the remaining 50 percent due upon receipt of a first-year interim report. As of January 2013, NASDARF had incurred and reported $44,492 for the subgrant under the previous EPA Agreement No. 83235401, and the remaining amount of $44,492 is reported under EPA Agreement No. 83456201.

We asked NASDARF to provide documentation of its review of the reasonableness of proposed costs for the award. In response, NASDARF stated that the costs were reasonable. NASDARF did not submit any comparisons of price quotes, market prices or other benchmarks to support its determination of reasonableness. In addition, NASDARF did not submit any documentation indicating that it had evaluated each element of cost.

CropLife Latin America Agreement

NASDARF made an award to CropLife Latin America for a project in Costa Rica. Dated January 31, 2011, the agreement included total project costs of $204,307 over a 3-year period. NASDARF's share of the total was $100,000. As of January 2013, NASDARF had incurred and reported the full $100,000.

We asked NASDARF for documentation of its review of the reasonableness of costs for the award. In response, NASDARF provided a document that described its review of the proposed amount. The analysis did not demonstrate that the proposed costs were reasonable. NASDARF stated that the cost of the project compared favorably to the cost of other EPA projects. NASDARF stated that the cost of the other EPA projects was $30,000. However, NASDARF did not name the other EPA projects, provide any supporting documents or other market prices, or provide benchmarks to support the reasonableness of the costs. NASDARF also stated that the project would be substantially more expensive if another entity conducted the project, but NASDARF did not provide any supporting documentation for this statement.

Written Procurement Policy Does Not Include Required Procedures

We reviewed NASDARF's written procurement policy to determine whether the policy included provisions required by 40 CFR Part 30. The policy contained several required procurement practices but lacked other requirements. NASDARF's policy was implemented in FY 2010 and applies to purchases exceeding $25,000. The policy:

- Prohibits conflicts of interest.
- Includes provisions for conducting and documenting a cost analysis in conjunction with every procurement.
- Requires justification for lack of competition if competitive bids or offers are not obtained.

- Requires procurements to be conducted to maximize opportunities, increase quality and reduce the cost of the purchase.
- Requires justification for the award of a contract to an entity other than the low bidder.

However, the NASDARF policy did not include the following requirements from 40 CFR Part 30, which requires solicitations to include:

- A clear statement on all requirements that the bidder or offeror should fulfill for the bid or offer to be evaluated (§30.43).
- A clear and accurate description of the technical requirements (§30.44(a)(3)(i)).
- Requirements that the bidder/offeror must fulfill all other factors to be used in evaluating bids or proposals (§30.44(a)(3)(ii)).
- Contracts be awarded only to responsible contractors who possess the potential ability to perform successfully under the terms and conditions of the proposed procurement (§30.44(d)).
- Procurement records and files for purchases in excess of the small purchase threshold also include the basis for contractor selection (§30.46).

Conclusion

NASDARF's procurements did not follow its own procurement policy, comply with federal requirements or meet all conditions of the agreement. When recipients do not justify sole-source procurements or complete the required cost or price analysis, the OIG has no assurance that costs are fair and reasonable. Since NASDARF did not document its decisions at the time of the procurement, it did not comply with federal requirements. Consideration of NASDARF's after-the-fact explanations and documentation is at the discretion of EPA management and would require a formal deviation from agency policy in accordance with 40 CFR §30.4, *Deviations*. This regulation states:

> The Office of Management and Budget (OMB) may grant exceptions for classes of grants or recipients subject to the requirements of Circular A–110 when exceptions are not prohibited by statute. However, in the interest of maximum uniformity, exceptions from the requirements of Circular A–110 shall be permitted only in unusual circumstances. EPA may apply more restrictive requirements to a class of recipients when approved by OMB. EPA may apply less restrictive requirements when awarding small awards, except for those requirements which are statutory. Exceptions on a case-by-case basis may also be made by EPA.

Recommendation

We recommend that the Director of the Office of Grants and Debarment:

5. Require NASDARF to establish controls for future awards in order to ensure:

    a. Documentation is maintained in procurement files to justify sole-source procurements and to ensure compliance with 40 CFR§30.46.
    b. Compliance with 40 CFR §30.45 by conducting a cost or price analysis to determine reasonableness of costs.
    c. Compliance with administrative conditions of the award by determining and documenting the reasonableness of subgrant costs.

NASDARF Response

NASDARF generally did not agree with the OIG findings and recommendations and provided the following comments.

Sole-Source Procurement Not Justified, Cost or Price Analysis Not Conducted: NASDARF responded that, as provided in previous communications with the OIG, both NASDARF and the EPA engaged several individuals for consideration to replace Mr. Richard Herrett as the NASDARF Project Manager. However, due to the unique skill-set and expertise required to adequately fill the NASDARF Project Manager position, only a finite number of individuals met the qualifications needed for consideration. One of the candidates, Mr. Roger Flashinski, under consideration for the Project Manager position, provided a statement confirming that he was not able to accept the position due to timing considerations and his position with another university did not allow for the same accommodations as Ms. Black. Additionally, NASDARF addressed the reasonableness of pay and provided comparison data to current positions similar in scope and demands.

Cost or Price Analysis Not Conducted: NASDARF did not agree with the OIG findings and recommendations and restated its assessment of "reasonableness of costs" for the University of Florida project. NASDARF stated that the rates in the proposal are reasonable and fall within typical rates for editing. Further, the pesticide-related subject matter is technical and requires a skilled editor. NADARF also included analysis of the various cost elements.

Similar to the University of Florida subgrant, NASDARF stated that it provided sufficient documentation to demonstrate "reasonableness of costs" for the CropLife Latin America project. NASDARF also included further discussion of the various cost elements and comparisons to other state pesticide safety education programs.

EPA Response

The EPA agreed with the OIG's recommendations and stated that OGD will require NASDARF to comply with the requirements of 40 CFR Part 30 and administrative conditions with respect to documenting the justification of sole-source procurements and performing an adequate cost or price analyses for procurements and subawards to determine the reasonableness of cost.

OIG Comments

The OIG acknowledges NASDARF's discussion regarding the cost-price analysis and reasonableness of costs associated with the Ramsey contract and the subgrants. However, since NASDARF did not document the analysis at the time of the procurement, it did not comply with federal requirements. Consideration of NASDARF's after-the-fact explanations and documentation is at the discretion of EPA management and would require a formal deviation from agency policy in accordance with 40 CFR §30.4, *Deviations*.

The OIG agrees with EPA's intended corrective actions for recommendation 5. However, because the EPA did not provide specific planned completion dates for the corrective actions, the OIG considers the status of the recommendation unresolved.

Chapter 6
Other Unresolved Issue

During our review of NASDA's single audit report for FY 2012, the OIG learned of an unresolved issue pertaining to a prior EPA agreement. Specifically, NASDARF reported and drew funds of $118,324 for potentially unallowable costs under EPA Agreement No. 83235401. Per the single audit report, the costs were recorded as a refundable advance and represent funds received as of year-end but not yet earned. NASDARF initially considered the costs (incurred in 2006 and 2007) as unallowable and did not report them. NASDARF later reported the costs to the EPA as part of the closeout of the agreement in 2011. Although the EPA closed the agreement, NASDARF stated that the agency never made a determination on the allowability of the costs. NASDARF will recognize these funds as revenue once accepted by the EPA.

We followed up with EPA personnel in the Office of Pesticide Programs and OGD, and found that staff were unaware of the issue. Personnel in both offices stated they had no knowledge of the potentially unallowable costs. However, staff said they would discuss the issue with personnel associated with the previous agreement and obtain more information. As a result, we questioned the $118,324 pending review and approval by EPA.

Although the agreement is closed, the EPA can recover funds if the funds are determined to be unallowable. Title 40 CFR §30.72(a) states the closeout of an award does not affect, in part, the following:

- The right of EPA to disallow costs and recover funds on the basis of a later audit or other review.
- The obligation of the recipient to return any funds due as a result of later refunds, corrections or other transactions.
- Audit requirements (§30.26).

In addition, 40 CFR §30.73(a) states:

> Any funds paid to a recipient in excess of the amount to which the recipient is finally determined to be entitled under the terms and conditions of the award constitute a debt to the Federal Government. If not paid within a reasonable period after the demand for payment, EPA may reduce the debt by paragraph (a) (1), (2) or (3) of this section.
>
> > (1) Making an administrative offset against other requests for reimbursements.

(2) Withholding advance payments otherwise due to the recipient.

(3) Taking other action permitted by statute.

Recommendation

We recommend that the Director of the Office of Grants and Debarment:

6. Determine the allowability of the $118,324 in costs incurred under prior EPA Agreement No. 83235401 and recover any costs determined to be unallowable.

NASDARF Response

NASDARF responded that the EPA had officially closed the grant and released the funds. NASDARF also provided a copy of the close-out letter.

EPA Response

The EPA responded to the recommendation stating that OGD and the Office of Chemical Safety and Pollution Prevention will require NASDARF to submit documentation for the costs in question incurred in 2007 under Agreement No. 83235401 to determine whether they are allowable under the agreement. EPA also responded it will review the documentation and take necessary corrective action, including the recovery of costs as appropriate if they are determined to be unallowable.

OIG Comments

The OIG acknowledges that the EPA closed the grant and released the funds. However, based on NASDARF's comments to the OIG and the single audit report for FY 2012, allowability of the costs remains at issue.

The OIG agrees with EPA's intended corrective actions for recommendation 6. However, because the EPA did not provide specific planned completion dates for the corrective actions, the OIG considers the status of the recommendation unresolved.

Status of Recommendations and Potential Monetary Benefits

		RECOMMENDATIONS				POTENTIAL MONETARY BENEFITS (in $000s)	
Rec. No.	Page No.	Subject	Status[1]	Action Official	Planned Completion Date	Claimed Amount	Agreed-To Amount
1	7	Disallow and recover $571,626 of questioned costs. If NASDARF provides documentation that meets appropriate federal requirements or demonstrates the fairness and reasonableness of the subcontract and subgrant costs, the amount to be recovered may be adjusted accordingly.	U	Director, Office of Grants and Debarment		$572	
2	15	Require NASDARF to: a. Recalculate its indirect cost rate for FY 2011, excluding subgrant and subcontract amounts in excess of $25,000; and submit to the NBC for approval, or ask the NBC to amend, the indirect cost rate agreements to include an equitable allocation base. b. Claim indirect costs using the recalculated approved rates	U	Director, Office of Grants and Debarment			
3	15	Require NASDARF to calculate its indirect cost rates for years beyond FY 2011 by excluding subgrant and subcontract amounts in excess of $25,000; or to be in accordance with any revised indirect cost rate agreement.	U	Director, Office of Grants and Debarment			
4	15	Require the following special conditions be included for all current and future awards until the EPA determines that NASDARF has met all applicable federal financial requirements: a. Payment on a reimbursement basis. b. Review and approval of reimbursement requests, including all supporting documentation, prior to payment.	U	Director, Office of Grants and Debarment			
5	23	Require NASDARF to establish controls for future awards in order to ensure: a. Documentation is maintained in procurement files to justify sole-source procurements and to ensure compliance with 40 CFR§30.46. b. Compliance with 40 CFR §30.45 by conducting a cost or price analysis to determine reasonableness of costs. c. Compliance with administrative conditions of the award by determining and documenting the reasonableness of subgrant costs.	U	Director, Office of Grants and Debarment			
6	26	Determine the allowability of the $118,324 in costs incurred under prior EPA Agreement No. 83235401 and recover any costs determined to be unallowable.	U	Director, Office of Grants and Debarment		$118	

[1] O = recommendation is open with agreed-to corrective actions pending
C = recommendation is closed with all agreed-to actions completed
U = recommendation is unresolved with resolution efforts in progress

NASDARF's Comments on Draft Report and OIG Responses

National Association of State Departments of Agriculture
4350 North Fairfax Drive
Suite 910
Arlington, VA 22203
Tel: 202-296-9680 | Fax: 703-880-0509
www.nasda.org

December 19, 2013

Ms. Angela Bennett
US EPA Region 4
Office of Inspector General
61 Forsyth Street, S.W.
Mailcode: 12T26
Atlanta, GA 30303-8960

> RE: Office of Inspector General's Draft Report on Examination of NASDARF's Cooperative Agreement No. 83456201

Dear Ms. Bennett:

Thank you for the opportunity to provide comments and additional information on the Office of Inspector General's (OIG) draft examination on the National Association of State Departments of Agriculture Research Foundation's (NASDARF) cooperative agreement with the Environmental Protection Agency (No. 83456201).

In an effort to provide a complete and accurate response to OIG's inquires under Project No. OA-FY13-0140, NASDARF has personally met with OIG management on several occasions, fully complied with all of OIG's requests for information and supporting documentation over several months, and met with EPA's Office of Pesticide Programs (OPP) representatives to discuss and provide informed responses to OIG's draft examination. In addition to the specific responses below, we are providing the attached documentation for your review (Attachments A-U).

We value OIG's role and mission, and we appreciate any opportunity to increase NASDARF efficiencies in our cooperative agreement with EPA. Please see our responses to OIG findings below.

OIG Charge: Financial Management Systems did not Meet Certain Federal Requirements

In Chapter 4 of the OIG draft examination, OIG made the following charges:

NASDARF's financial management system did not meet certain federal regulations and agreement conditions. Specifically, NASDARF:

- *Did not calculate and apply indirect cost rates correctly.*
- *Reported outlays for indirect costs in excess of recorded expenses in the general ledger.*
- *Drew down funds that exceeded cash needs.*

These conditions occurred because NASDARF did not have controls in place to ensure compliance with federal regulations and agreement conditions. In calculating indirect cost rates, NASDARF was not aware initially of 2 CFR Part 230 requirements to exclude subgrants and subcontracts in excess of $25,000. NASDARF later interpreted the requirements erroneously and only excluded amounts for subgrants, but not subcontracts. For the reported outlays, NASDARF lacked controls to ensure the recording of adjustments in the general ledger. Lastly, NASDARF's written procedures for drawdowns did not incorporate cash-management requirements as described in the administrative conditions of the agreement or under federal regulations.

NASDARF Response to Chapter 4 Charge: *Did not calculate and apply indirect cost rates correctly:*

NASDARF is fully informed of 2 CFR Part 230 requirements to exclude subgrants and subcontracts in excess of $25,000, and at no time was NASDARF under the impression, nor did NASDARF execute any activities under the illusion, that indirect costs could be charged on subgrants and subcontracts in excess of $25,000.

OIG's charge relates specifically to Ms. Carol (Ramsay) Black's misclassification as a "subcontractor," when in fact, Ms. Black is a contracted employee/consultant under Ramsay Consulting, LLC. NASDARF maintains a long standing relationship with Ms. Black in her capacity as a contract employee, and Ms. Black's correct classification as a contracted employee is not subject to the rule of a subcontract under 2 CFR Part 230.

As provided in previous communications with OIG, both NASDARF and EPA engaged several individuals for consideration to replace Mr. Richard Herrett as the NASDARF Project Manager. However, due to the unique skill-set and expertise required to adequately fill the NASDARF Project Manager position, only a finite number of individuals met the qualifications needed for consideration. Subsequently, both EPA and NASDARF identified Ms. Black as the most qualified employee candidate during the recruitment process to replace Mr. Herrett, who had indicated his desire to retire. However, in order to hire Ms. Black and allow her to retain her faculty position at Washington State University, it was necessary for Ms. Black to serve in a "consultant/employment" capacity. To this end, NASDARF made a budgetary request in March 2011 to move $520,000 from the Salary budget category into the Contractual category.

Since Ms. Black serves as an agent for NASDARF and the project manager for cooperative agreement No. 83456201, her position should be treated as direct salary. NASDARF recognizes Ms. Black's hourly expenses should be budgeted as salary per the original budget. NASDARF will request that the budget is returned to its original classifications.

Ms. Black is perceived by all project partners, stakeholders, and EPA as being an agent of NASDARF. Her expertise is largely responsible for the high level of quality work and numerous deliverables this Cooperative Agreement has produced, and EPA, NASDA, and the stakeholder community hold her in high regard. OIG and NASDARF may disagree on Ms. Black's position and the Ramsay Consulting, LLC contract. Given this potential disagreement, NASDARF maintains the indirect cost rule was applied consistently across each and every other subgrant and subcontract, including those in question with University of Florida, CropLife Latin America, and Texas A&M University.

In addition, OIG listed Ramsay Consulting LLC, Richard Herrett, Marla Stein Associates, and CropLife Latin America as subcontractors/subgrantors receiving payments in excess of $25,000. Ramsay Consulting, LLC and Mr. Herrett served as project managers for the grant and are therefore considered direct salary. Irrespective of the classification, Mr. Herrett was not paid under this current agreement. Marla Stein Associates received $21,143 under Cooperative Agreement X8-83456201; these expenses are as follows: printing Handler Verification cards (second payment of $11,950 from June 2010), a thumb drive ($163 in Sept 2012) and publication design for a PPE document ($9,030 in May 2013). The remaining invoices were charged under a prior agreement (Attachments A-H).

As OIG noted in its discussion of adjustments to the indirect allocation, indirect costs were not ultimately charged on the $10,093 over $25,000 paid to CropLife Latin America.

NASDARF applies for their indirect rate annually. The rate is approved by the National Business Center (NBC). This results in a revised Final indirect rate and a new Provisional indirect rate. The Final indirect rate is received long after the close of the NASDARF fiscal year and almost always requires adjustment to prior years (Attachment I).

NASDARF is awaiting the EPA grants office's opinion before making any decisions in regards to reapplying for indirect rates. Indirect rates were calculated with guidance from EPA and the NBC, which considered all programs currently undertaken by NASDA and NASDARF. This arduous process takes resources away from program goals and would more than likely result in no material adjustments.

OIG Response 1: In its response, NASDARF states that Ms. Black's position should be treated as direct salary. The OIG disagrees with this position. Under 2 CFR 230, Appendix B.8, compensation for personal services includes services provided by the organization's employees. Since Ms. Black is not an employee, her position cannot be treated as a direct salary. There is no disagreement that Ms. Black is a contractor. However, based on further review of 2 CFR Part 230, the OIG believes that clarification is needed about the relationships discussed (i.e., grants-subgrants, contracts-subcontracts, and whether the requirement applies to the Ramsey contract). NASDARF's relationship with Ramsey is not the typical grant-subgrant relationship; rather, it is a grant-contract relationship. We recommend that the EPA seek clarification on whether this type of relationship is subject to the $25,000 requirement. This would also apply to Mr. Herrett.

NASDARF Response to Chapter 4 Charge: *NASDARF lacked controls to ensure the recording of adjustments in the general ledger:*

NASDARF disagrees with OIG's statement that NASDARF lacked "controls to ensure the recording of adjustments in the general ledger." The nature of any indirect rate application results in regular adjustments when rates change and new payments are made for subgrants. NASDARF's goal is to obtain a correct cumulative total in the indirect allocation, not an unrealistic expectation of an unchanging allocation each month. Assessing these values at interim periods over the life of the grant only offers a snapshot at that specific time. Attachment J includes the general ledger through October 2013 and Attachment K is a revised SF425 as of October 31, 2013, which provide greater context of the grant during a dynamic, interim period. Attachment L reflects the indirect rates over the life of the grant.

> **OIG Response 2:** The OIG based its finding on the fact that the general ledger did not support the outlays reported on the interim FFR. The OIG acknowledges the adjustments of indirect costs as NASDARF discussed; however, outlays reported on the FFR should be adequately supported as required by 40 CFR §30.21(b)(7). NASDARF did not provide any additional documentation to support the two adjustments reported under the interim FFR. As a result, the OIG continues to question the costs as unsupported.

NASDARF Response to Chapter 4 Charge: *NASDARF's written procedures for drawdowns did not incorporate cash-management requirements described in the administrative conditions of the agreement or under federal regulations:*

NASDARF has changed both personnel and policy within the last year to insure cash management practices are in line with grant requirements. According to the revised SF425, as of October 31, 2013, EPA owed NASDARF $125,319.20 in direct expenses and indirect allocations (Attachment M). This money was drawn down in November per NASDARF's updated policy, which now reads:

> "Drawdowns related to the EPA agreement are initiated by the Director of Finance based on anticipated short-term expenses as reported and documented by the Project Director. Drawdowns are typically made after expenses are paid but may be made a week in advance in anticipation of large vendor/subgrant disbursements."

NASDARF is aware of OIG's position claiming funds are available within 3 business days; however, recent technical difficulties with the ASAP website make it extremely challenging for NASDARF management to operate with a 3-day window for distributing funds.

> **OIG Response 3:** The OIG acknowledges that NASDARF revised its procedures; however, the procedures do not indicate that drawdowns should be made as close as administratively feasible to the actual disbursements and be timed in accordance with the actual, immediate cash requirements. The OIG does not consider this acceptable, and continues to maintain that NASDARF can get funds from the EPA within 3 days or less.

NASDARF Response to Chapter 4 Charge: *NASDARF used EPA funds to cover expenditures under other non-EPA programs:*

OIG's assertion that NASDARF used EPA funds for expenditures for other non-EPA programs is incorrect. Funds were transferred from the NASDARF checking account to other NASDA affiliated accounts only after NASDARF earned the funds through indirect allocations or reimbursement for direct expenses. NASDARF is unclear how OIG's assumption was generated, and NASDARF is highly concerned OIG did not adequately develop an understanding of NASDARF's revenue recognition and cash management procedures when reaching this conclusion. NASDARF welcomes additional information and explanation on how OIG developed this finding.

OIG Response 4: The OIG disagrees with NASDARF that it did not use EPA funds for expenditures for non-EPA programs. The OIG's cash flow analysis, conducted using NASDARF's bank statements, showed that NASDARF moved funds between the EPA's programs and others. Most of the transfers out were to the NASDA account, which, as discussed in the response above, could be for reimbursement for indirect expenses. However, there were significant deposits to and from the EPA account from other agencies. There were also smaller transfers out from the EPA account to other associations within the state departments of agriculture. Title 40 CFR 30.21(b)(3) requires that recipients' financial management systems shall assure that funds are used solely for authorized purposes.

OIG Charge: Procurements did not meet Federal Requirements or Agreement Conditions

In Chapter 5 of the OIG draft examination, OIG made the following charges:

> *NASDARF's procurement practices for subcontracts and subgrants did not comply with federal requirements or agreement conditions. In particular, NASDARF did not:*
>
> - *Provide documentation to support the sole-source procurement process used for its project management subcontract, or conduct a cost or price analysis as required by 40 CFR Part 30 and NASDARF's own procurement policy.*
> - *Comply with agreement administrative condition no. 15.a.(4) for subgrants awarded to the University of Florida and CropLife Latin America.*
> - *Include provisions required by 40 CFR Part 30 in its written procurement policy.*
>
> *These conditions occurred because of NASDARF's lack of knowledge or a lack of understanding about agreement conditions and federal requirements. Additionally, NASDARF's written policies and procedures provided inadequate guidance for subcontract and subgrant awards. As shown in chapter 3, we questioned $295,976 that NASDARF incurred and reported for three awards. – not explicitly addressed.*

NASDARF Response to Chapter 5 Charge: *Sole-Source Procurement Not Justified; Cost or Price Analysis Not Conducted:*

As stated above, Ms. Black is a contracted employee/consultant under Ramsay Consulting, LLC. NASDARF maintains a long standing relationship with Ms. Black in her capacity as a contract employee, and Ms. Black's correct classification as a contracted employee is not subject to the rule of a subcontract. This was an employee hiring decision for which both NASDARF and EPA Office of Pesticide Programs ("OPP") participated.

As provided in previous communications with OIG, both NASDARF and EPA engaged several individuals for consideration to replace Mr. Richard Herrett as the NASDARF Project Manager. However, due to the unique skill-set and expertise required to adequately fill the NASDARF Project Manager position, only a finite number of individuals met the qualifications needed for consideration. One of the candidates, Mr. Roger Flashinski, under consideration for the Project Manager position provided a statement confirming he was not able to accept the position due to timing considerations and his position with another university did not allow for the same accommodations as Ms. Black (Attachment N).

In response to the issue of reasonableness of pay, the project manager position requires a skill set similar to those of a government employee program manager or branch chief. NASDARF's project manager is responsible for overseeing multiple programs areas, including: applicator certification, worker protection handler education and support for health care providers. A qualified project manager must have demonstrated extensive national leadership and be highly respected in pesticide applicator education and regulation arena. The manager must be well-established within the wide network of national stakeholders and be well-respected within the EPA Office of Pesticide Programs. They must have demonstrated success in coalition leadership, conference management, project development & coordination, and manage a highly respected pesticide safety education/regulatory program.

NASDARF based Ms. Black's hourly compensation on similar positions located in the Washington, DC and Seattle, WA areas. Furthermore, NASDARF discussed Ms. Black's pay level with EPA OPP, which subsequently agreed on appropriateness of Ms. Black's pay level. Current positions similar in scope and demands include:

- National Program Leader for IR-4 and PMP at USDA National Institute for Food and Agriculture (NIFA) (Monte P. Johnson). Mr. Johnson provides national leadership for state and federal activities aimed at developing a greater understanding of the toxicological consequences of human exposure to pesticides and the effects of pesticide residues in foods and the environment. Prior to assuming this position with USDA, Mr. Johnson was the University of Kentucky pesticide education specialist. Current rank: GS-15 and salary: $148,510 (or $71.40/hour), not including benefits.
- USDA Plant Systems Protection Program Leader (Mary Purcell-Miramontes). Ms. Purcell-Miramontes leads and directs Agriculture & Food Research Initiative (AFRI) competitive grant programs on Insects and Nematodes in Plant Systems and co-chairs the Colony Collapse Disorder Steering Committee and Program Leader for the Pesticide Safety Education Program. Current rank: GS-15 and salary: $140,259 (or $67.43/hour), not including benefits.

- Branch Chief Worker Safety Division for US Environmental Protection Agency (Kevin Keaney). Mr. Keaney works for OPP's Field and External Affairs Division and manages regional and state coordination and assistance for applicator certification, worker protection, health incidents, and health care providers, including communication and outreach activities. Current rank: GS-15 and salary: $155,500 (or $74.52/hour), not including benefits.
- USDA NIFA Office of the Director, Program and Analysis Officer (William J. Hoffman). Mr. Hoffman participates in programming, policy development and interpretation, development and updating of NIFA's guidelines, COOP plans, and resources. Current rank: GS-14 and salary: $126,251 (or $60.70/hour), not including benefits.

Cost basis for GS Pay Scale for positions in Seattle, WA and Washington, DC:
- GS 14 -- $65.53 per hour + 37% benefit rate = $89.78
- GS 15 – 74.51 per hour + 22.78% benefit rate = $91.48 (actual benefit rate for EPA GS15 is 22.78%)

Ms. Black's contract with NASDA states the $90.00/hour includes a base pay rate of $74.50 per hour and $15.50 per hour overhead expenses for the contractor. The base rate would actually be higher had Ms. Black been a salaried employee for NASDARF. The employee benefits rate at NASDA for retirement, health, and dental is 24.5%. The benefit rate for federal employees ranges from 23-37%. Ms. Black's rate for benefits is currently 20%.

OIG Response 5: The OIG acknowledges the NASDARF's discussion regarding the selection of Ms. Ramsey. However, since NASDARF did not document its decisions at the time of the procurement, it did not comply with federal requirements. Consideration of NASDARF's after-the-fact explanations and documentation is at the discretion of EPA management and would require a formal deviation from agency policy in accordance with 40 CFR §30.4, *Deviations*.

NASDARF Response to Chapter 5 Charge: *Cost or Price Analysis Not Conducted:*

NASDARF restates its assessment of "reasonableness of costs" for the UFL project. We disagree with OIG finding our assessment did not measure a prudent person's assessment. The rates in the proposal are reasonable and fall within typical rates for editing as referenced by the Editorial Freelancers' Association (http://www.the-efa.org/res/rates.php), ProComm (Attachment O) and HOW Design (http://www.howdesign.com/design-business/pricing/hourly-rates). The pesticide-related subject matter (air sampling, data analysis) is technical and requires a skilled editor. Presentation development includes content development, graphic design and multimedia animation and is priced similarly. Narration includes the voice over narration, plus insertion and synchronization with the base media (PowerPoint or video). Using our collective experience on the amount of time to produce publications and current market prices, the below description is accurate:

Cost Analysis:

- OTHER $13,000 – publication and postage way under budget:
  o When looking at expenses for curriculum development, web design, print design, presentation design, which includes technical experts, editors, designers, professional

narration, total costs are reasonable and anticipate an outlay for publication and presentation design to be over $2,500.

- Editors $1,000 – range in per/hour fees: 800-1600 words per hour. $50-125/hour depending on level of content expertise required.
- Publication Designers $1,000 - $45-125 per page – anticipate 10 pages
- Presentation development – 40 hours at $45-100/hour
- Professional narration $500 - $500 per completed hour (prep, narrate, edit)
  - o Printing/mailing costs for follow-up survey and color-brochures are reasonable
    - Follow-up survey mailed to each school district (3,928 copies) – print $4,000 ($1/copy), send and return envelopes $800 = $4,800
    - Tri-fold color brochure – estimate 4,000 copies at $1 each - $4,000
  - o Postage ($1.20/envelope) = $4,714 per mailing
- SALARIES and WAGES $13,547:
  - o Personnel: this project is receiving more commitment of time than what was sought in the budget
    - 0.04 FTE equates to 83 hours or just over two-weeks time for a lead coordinator and assistant coordinator. This project will incur many more hours than estimated. It is anticipated that a time commitment of 0.25 FTE would be more appropriate. The base salary is $145,000 for the Extension Specialist (GS15 equivalent) and $116,250 (GS13 equivalent) for the Assistant position. These are equitable salary ranges for senior university extension personnel when compared to federal employee pay scales.
    - Benefits are listed at 27.8% which covers health insurance premiums, dental, life insurance premiums and Florida Retirement System Pension Plan and Investment Plan.
- TRAVEL $9,360: Travel: meets state requirements for per diem and least expensive travel options; believe this is under-budgeted for anticipated expenses
  - o Estimate 2 interstate trips for 2 people (airfare, per diem) = $1,200 per trip x 4 = $4,800
  - o Estimate 6 interstate trips for 2 people (mileage, per diem) = $440 per trip x 12 = $5,280

Similar to the University of Florida subgrant, NASDARF provided sufficient documentation to demonstrate "reasonableness of costs" for the CropLife Latin America, Costa Rica project. We refer OIG and EPA to our original response. CropLife Latin American project effort continues EPA's presence in supporting Central American pesticide safety education programs (El Salvador, Honduras, Costa Rica, Nicaragua) (Attachment P and Q).

There is significant financial benefit with partners providing matching expenses: FLNC & their partners will provide financial and substantial in-kind support, CLLA will provide $50,000 and will provide oversight and accountability reporting for the program.

Personnel: CLLA and FLNC will provide personnel with expertise in implementing pesticide safety train-the trainer programs and coordinating diverse stakeholder coalitions; expect to have additional expert support from: Costa Rica Ministries of Agriculture, Labor, Environment, and Health; pesticide manufacturers industry; University of Costa Rica; International Labor Organization; Instituto Nacional de Aprendizaje (national training organization); Cooperativa de Productores de Leche Dos Pinos (milk producers association); and other agricultural producer cooperatives.

The reach of the outreach program includes farmers, managing applicators, pesticide handlers, dealers, and field workers.

The project would be substantially more expensive if another entity conducted it because another entity would need to expend time and resources to develop the strong connections with a broad group of stakeholders and an understanding of the needs of pesticide applicators and other people involved in pesticide sales, use, and disposal. Having the experience from other Latin American countries is invaluable.

To augment the response, we provide the following information:

The one-year report (June 2004-2005) and budget from the 2003 CropLife Honduras project shows a very similar scope of work, budgetary items and budgeting levels. The one-year budget which trained over 8,000 people was nearly $48,000 (Salaries: $14,000, Supplies/Training Aids: $2,100, Equipment: $4,800, Printed Materials: $4,600, Marketing/Communication: $ 2,000, and Travel $18,800) (Attachment R).

NASDARF also compared what EPA/USDA typically provides to each well-established U.S. state certified applicator education program ($30,000 base funds + additional funds based on the number of certified applicators, requires 100% matching funds from the state); the proposed annual cost of the Costa Rica effort (approx. $33,333 per year from NASDARF with matching funds from CLLA and FLNC, as well as substantial in-kind support) and the proposed implementation costs appear reasonable. Below are samples of four states from each region of the United States for the USDA Budget year FY2008. The on the next page table indicates the total annual support level provided by the EPA/USDA Funding and State Match. This total excludes agent and specialist salaries, benefits and operations. The last column indicates the percentage of the total operating budget supported by EPA/USDA. The Costa Rica project fits well within this range (Attachments S and T).

State Pesticide Safety Education Programs (PSEP) Note: these are established programs. Source: Data from USDA base-support in FY2008 allocation; funding originated from EPA				approx. % of total PSEP operations
State	**USDA Funding**	**University Match**	**Total**	
Southern Region				
University of Georgia	$42,668	$42,668	$85,336	100%
University of Arkansas	$34,107	$34,107	$68,214	50%
Oklahoma State University	$38,164	$38,164	$76,328	25%
North Carolina State University	$48,766	$48,766	$97,532	40%
North Central Region				
Iowa State University	$47,142	$47,142	$94,284	20%
Kansas State University	$32,696	$32,696	$65,392	100%
Michigan State University	$39,701	$39,701	$79,402	100%
North Dakota State University	$29,841	$29,841	$59,682	25%
Northeast Region				
Virginia Polytechnic Institute & State University	$28,925	$28,925	$57,850	15%
University of Massachusetts	$19,166	$19,166	$38,332	35%
University of Maine	$18,003	$18,003	$36,006	33%
Pennsylvania State University	$47,169	$47,169	$94,338	15%
Western Region				
Colorado State University	$26,798	$26,798	$53,596	45%
University of Alaska	$16,898	$16,898	$33,796	75%
University of Wyoming	$33,474	$33,474	$66,948	100%
Washington State University	$43,083	$43,083	$86,166	15%

NASDARF also directs OIG to the Office of Grants and Debarment opinion related to the procurement procedures for the University of Florida and CropLife Latin America awards, given the controls already in place for awards of this type.

OIG Response 6: The OIG acknowledges NASDARF's discussion regarding the cost-price analysis and reasonableness of costs associated with the subgrants. However, since NASDARF did not document the analysis at the time of the procurement, it did not comply with federal requirements. Consideration of NASDARF's after-the-fact explanations and documentation is at the discretion of EPA management and would require a formal deviation from agency policy in accordance with 40 CFR §30.4, *Deviations*.

OIG Charge: Other Unresolved Issue

In Chapter 6 of the OIG draft examination, OIG made the following charges:

> *During our review of NASDA's single-audit report for FY 2012, the OIG learned of an unresolved issue pertaining to a prior EPA agreement. Specifically, NASDARF reported and*

drew funds of $118,324 for potentially unallowable costs under EPA Agreement No. 83235401. Per the single-audit report, the costs were recorded as a refundable advance and represent funds received as of year-end but not yet earned. NASDARF initially considered the costs (incurred in 2006 and 2007) as unallowable and did not report them. NASDARF later reported the costs to the EPA as part of the closeout of the agreement in 2011. Although the EPA closed the agreement, NASDARF stated that the agency never made a determination on the allowability of the costs. NASDARF will recognize these funds as revenue once accepted by the EPA.

NASDARF Response to Chapter 6:

NASDARF has attached the letter officially closing this grant and funds have been released (Attachment U).

OIG Response 7: The OIG acknowledges that EPA closed the grant and released the funds. However, based on NASDARF's comments to the OIG and the single audit report for FY 2012, allowability of the costs remains at issue. On this basis, the OIG continues to question the cost. The EPA should review and determine the allowability of the costs and recover any determined to be unallowable.

Conclusion

NASDARF appreciates OIG's efforts and in response to these OIG findings, NASDARF modified our subcontract for project management services and updated our written procurement procedures to include OIG-recommended requirements pertaining to 40 CFR Part 30.

NASDARF is pleased to partner with EPA in providing these timely, lifesaving programs to pesticide applicators around the U.S. and Central America. NASDARF will continue to work with the Office of Pesticide Programs and the Office of Grants and Debarment to achieve the goals of the program while maintaining a commitment to excellence.

We appreciate any opportunity to increase NASDARF efficiencies in our cooperative agreement with EPA. We stand ready to work with our federal partners and OIG to continue to improve these processes. Please let us know if you have any questions or would like to discuss further.

Sincerely,

Stephen Haterius

Stephen Haterius
Chief Executive Officer, NASDA

EPA's Comments on Draft Report

UNITED STATES ENVIRONMENTAL PROTECTION AGENCY
WASHINGTON, D.C. 20460

OFFICE OF
ADMINISTRATION
AND RESOURCES
MANAGEMENT

December 20, 2013

MEMORANDUM

SUBJECT: Response to Office of Inspector General Draft Report, Project No. OA-FY13-0140 titled, " National Association of State Departments of Agriculture Research Foundation (NASDARF) did not comply with certain Federal requirements and EPA award conditions", dated November 5, 2013

FROM: Kysha Holliday
Deputy Director, NPTCD

TO: Robert K. Adachi.
Director of Forensic Audits
Office of the Inspector General

Thank you for the opportunity to respond to the issues and recommendations in the subject draft audit report. Attached is a summary of the Office of Grants and Debarment's (OGD) and the Office of Chemical Safety and Pollution Prevention's (OCSPP) overall position regarding each report recommendation.

AGENCY'S OVERALL POSITION

OGD and OCSPP generally agree with the Inspector General's findings that NASDARF did not comply with certain requirements of 40 CFR Part 30, 2 CFR Part 230, or its award conditions.

As described below, OGD and OCSPP agree with Recommendations 1 through 6.

AGENCY'S RESPONSE TO REPORT RECOMMENDATIONS

Agreements

No.	Recommendation	Intended Corrective Action(s)	Estimated Completion by FY
1	Disallow and recover $571,626 of questioned costs. If NASDARF provides documentation that meets appropriate federal requirements or demonstrates the fairness and reasonableness of the subcontract and subgrant costs, the amount to be recovered may be adjusted accordingly.	1. OGD will provide NASDARF the opportunity to submit documentation to substantiate the questioned costs. OGD will review the documentation and take necessary corrective action, including the recovery of all or part of the questioned subcontract and indirect costs as well as funds drawn. OGD will work with NASDARF to implement corrective actions to comply with federal requirements on assuring the reasonableness of sub-grants, sub-contracts, indirect costs and drawdown amounts.	Within 180 days of OIG issuing their final report or as soon as practicable
2	Require NASDARF to: a. Recalculate its indirect cost rate for FY 2011, excluding sub-grant and subcontract amounts in excess of $25,000; and submit to the NBC for approval, or ask the NBC to amend, the indirect cost rate agreements to include an equitable allocation base. b. Claim indirect costs using the recalculated approved rates.	2. OGD will provide NASDARF the opportunity to clarify its position and properly support its indirect rate and require NASDARF to amend the rate where appropriate. OGD will review NASDARF's support and coordinate with DOI-NBC to review and amend the ICR agreement as needed, then require NASDARF to adjust claimed costs according to the revised agreement.	Within 180 days of OIG issuing their final report or as soon as practicable
3	Require NASDARF to calculate its indirect cost rates for years beyond FY 2011 by excluding subgrant and subcontract amounts in excess of $25,000; or to be in accordance with any revised indirect cost rate agreement.	3. Where necessary, OGD will require NASDARF to recalculate their indirect costs rates to be in compliance with 2 CFR 230 and coordinate with DOI-NBC to revise NASDARF's rates. OGD will then require NASDARF to adjust claimed costs according to any revised ICR agreements.	tbd

No.	Recommendation	Intended Corrective Action(s)	Estimated Completion by FY
4	Require the following special conditions be included for all current and future EPA awards until the EPA determines that NASDARF has met all applicable federal financial requirements: a. Payment on a reimbursement basis. b. Review and approval of reimbursement requests, including all supporting documentation, prior to payment.	4. OGD will place NASDARF on reimbursement for their active assistance agreements. The Agency will require NASDARF to provide supporting documents for costs incurred for the Agency's review prior to releasing funds for payment. NASDARF will remain on reimbursement status until they have demonstrated to the Agency that their financial management of EPA agreements meets applicable federal requirements for drawing grant funds.	Within 180 days of OIG issuing their final report or as soon as practicable (ongoing)
5	Require NASDARF to establish controls for future awards in order to ensure: a. Documentation is maintained in procurement files to justify sole-source procurements and to ensure compliance with 40 CFR §30.46. b. Compliance with 40 CFR §30.45 by conducting a cost or price analysis to determine reasonableness of costs. c. Compliance with administrative conditions of the award by determining and documenting the reasonableness of subgrant costs.	5. OGD will require NASDARF to comply with the requirements of 40 CFR 30 and administrative conditions with respect to documenting the justification of sole source procurements and performing an adequate cost or price analyses for procurements and sub awards to determine the reasonableness of cost.	Within 180 days of OIG issuing their final report or as soon as practicable
6	Determine the allowability of $118,324 of costs incurred under prior EPA Agreement No. 83235401 and recover any costs determined to be unallowable.	6. OGD and OCSPP will require NASDARF to submit documentation for the costs in question incurred in 2007 under grant No. 83235401 to determine if they are allowable under the agreement. OGD and OCSPP will review the documentation and take necessary corrective action, including the recovery of costs as appropriate if they are determined to be unallowed.	

OGD and OCSPP have a valued relationship with NASDARF and an obligation to manage grants in accordance with federal fiduciary and stewardship standards. The Agency fully intends to take the necessary corrective actions and work with NASDARF to resolve the findings of the OIG audit once formally issued.

CONTACT INFORMATION
If you have any questions regarding this response, please contact Kysha Holliday, Deputy Director of NPTCD at (202)564-1639 or Joe Lucia (202) 564-5378.

cc:
        Arthur A. Elkins Jr., Inspector General
        Angela Bennett, Project Manager, Office of the Inspector General
        Howard Corcoran, Office of Grants and Debarment
        Carolyn Schroeder, OCSPP, OPP-FEAD
        Deborah Hartman, OCSPP, OPP-FEAD
        Kevin Keaney, OCSPP, OPP-FEAD
        Jill Young, GIAMD
        Denise Polk, Director NPTCD
        Barbara Proctor, GIAMD-AAO
        Kristen Arel, GIAMD
        Bernadette Dunn, OCFO

bcc: none

Distribution

Director, Office of Grants and Debarment, Office of Administration and Resources Management
Agency Follow-Up Official (the CFO)
Agency Follow-Up Coordinator
Associate Administrator for Congressional and Intergovernmental Relations
Associate Administrator for External Affairs and Environmental Education
Director, Grants and Interagency Agreements Management Division, Office of Administration
    and Resources Management
Audit Follow-Up Coordinator, Office of Administration and Resources Management
Audit Follow-Up Coordinator, Office of Grants and Debarment, Office of Administration
    and Resources Management

www.ingramcontent.com/pod-product-compliance
Lightning Source LLC
Chambersburg PA
CBHW081234170526
45165CB00009B/3055